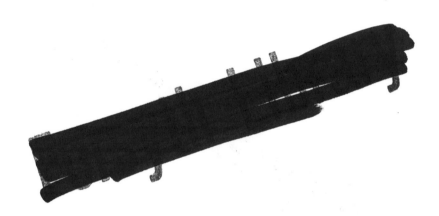

WITHDRAWN

PROFESSIONAL VIDEO GRAPHIC DESIGN

PROFESSIONAL VIDEO GRAPHIC DESIGN

Ben Blank
and
Mario Garcia

Prentice Hall Press • New York

Copyright © 1986 Prentice-Hall, Inc.
All rights reserved
including the right of reproduction
in whole or in part in any form

Published by Prentice Hall Press
A Division of Simon & Schuster, Inc.

PRENTICE HALL PRESS is a trademark of Simon & Schuster, Inc.

Library of Congress Cataloging-in-Publication Data

Blank, Ben.
  Professional video graphic design.

  1. Television graphics.   I. Garcia, Mario R.
II. Title.
PN1992.8.G7B53   1986   778.59   85-31250
ISBN 0-13-725797-X

Manufactured in the United States of America

10 9 8 7 6 5 4 3 2 1

CONTENTS

	Acknowledgments	vii
	Preface	ix
Chapter One	Video Graphic Design	1
	The Impact of Video Graphic Design	3
	Displayed Material	5
	Sound and Movement	7
	Symbolic Thinking	9
Chapter Two	Designing for the Screen	23
	Effective Design	25
	Perspective	33
	Spatial Distribution	35
	Clarity of Meaning	37
	The Designer and His Craft	39
Chapter Three	Typography for the Screen	47
	Purpose of Type	51
	Criteria for Selecting Type	55
	Placement of Type	58
	Creativity and Type	60
	Typography and Special Events	65
	Typographic Symbols	69

Chapter Four	Photographic Design for Television	71
	Criteria for Selecting a Still Photograph for Television	74
	Strategies with Still Photographs	76
	Doctoring Photographs for the Screen	81
	Developing and Maintaining Photographic Files	85
Chapter Five	Art and Illustration for the Screen	87
	When Should Illustrations Be Used?	90
Chapter Six	The Design and Use of Maps as Informational Graphics	97
	Characteristics	99
Chapter Seven	The Graphics of Major Anticipated News Events	107
	Elections	109
	Space Shots	114
	Olympic Games	116
Chapter Eight	Production Tips for the Video Graphic Designer	119
Chapter Nine	Computer-generated Graphics and the Machines That Produce Them	129
	Computer-generated Graphics	131
	About the Authors	157

ACKNOWLEDGMENTS

We could not have compiled this text without the assistance of many dedicated professionals who gave their time, talents, and experience to enrich the quality of the material presented here. We would like to thank Elmer Lower for his encouragement and for introducing us at a meeting in the fall of 1978 at the Syracuse University campus. What began as an interesting discussion over lunch became a four-year project that brought us together at many think sessions in New York City and in St. Petersburg, Florida. From the beginning, we were encouraged by colleagues at ABC News and at Syracuse University's Newhouse School of Public Communications.

Special thanks go to Jerry Cappa, art director of WLS-TV and president of the Broadcast Designers Association (BDA), to members of the BDA who graciously allowed their work to be used as illustrations for this book, to Zaro Calabrese and Louis Castellar of the ABC News Graphics Department for organizing much of the graphics used, to Gerald Andrea for his illustrations, and to Phil Cirrone for his maps. We also thank Allan Eastman of WCBS News Graphics for his design contributions; Ned Steinberg, graphics director of CBS News; and Gil Cowley, graphics director of WCBS News. John R. Malloy, managing editor for electronic products, Tribune Media Services, Orlando, Florida, contributed valuable data.

We are also grateful to Professors Peter Moller and David

Hollenback of the Telecommunications Department, S. I. Newhouse School of Public Communications, Syracuse University, and to Dean Edward Stephens and former dean Henry F. Schulte. Thanks also go to Robert Haiman, director of The Poynter Institute for Media Studies in St. Petersburg, Florida.

Special thanks to Bob Deramo, Ben Blank's secretary, for typing and correcting his copy.

Finally, this book would not have been possible without the patience and encouragement of our families.

PREFACE

Forty years ago television was an unheard-of phenomenon. Today it has changed the way in which millions of people are informed, entertained, and educated. It ranks high as one of the most influential developments of the twentieth century. And although its impact has been mostly favorable, critics of television claim that the medium has created a generation of nonreaders, and maintain that it has also cultivated isolation within the family nucleus by absorbing attention of family members in such a way as to eliminate conversation and the sharing of activities.

Although its harshest critics see the medium as an evil influence and a drain of valuable time for individuals, defenders of television describe it as one of history's most effective educators and a provider of escape from the troubles of daily life.[1] Whether one defends it or attacks it, the medium of television has undoubtedly affected the way in which our generation thinks; in some ways it shapes the lifestyles of many people. Sandman, Rubin, and Sachsman aptly describe television as a creator and perpetuator of mass culture, adding that America is by all counts the world's greatest producer of mass culture.[2]

The subject of television and its impact on the public is not limited to trade journals. Such diversified magazines as *Human Behavior, Psychology Today, The New York Times Maga-*

Frame A. Titling devices for the opening of a film or TV show on evolution versus creation.

zine, *The New Republic, Time,* and *Journalism Quarterly* devote articles to television, sometimes analyzing the effect television has on viewers, condemning its abuse of topics dealing with sex and violence, or praising the value of educational programs and good on-the-spot reporting. Most of the articles describe television viewing as an absorbing experience. And, indeed, the combination of audio and visual elements on a screen—the representation of visual elements to transmit information, to entertain, or to transport the viewer to faraway places—constitutes an *absorbing experience.*

Although the process of watching television requires little effort on the part of the viewer, it imposes tremendous demands on those in charge of producing the messages. As a medium affected by film, theatre, journalism, and technology, television needs the most talented, best trained, and most creative people for its impact.

When we originally conceived the idea for this book, we intended to orient it toward the creation of graphics for television news. The organization of the book underwent dramatic transformation as a result of such developments as cable television, the wider use of computer graphics within business and industry, and the more prevalent use of television as a medium in educational and nonbroadcast situations.

We firmly believe that the most functional and effective principles of designing for the *screen* are common to all video systems, including film. This book is aimed at persons involved in or being trained in the visual aspects of television

Frame B. Polish strike frame.

Frame C. Courtroom illustration (sketch).

Frame D. Economy/inflation frame.

Frame E. Solar energy frame.

Frame F. The world of video graphic design has expanded to include application in such new fields as "advertising on the air" through twenty-four-hour services blending news, entertainment, and ads, as shown through this frame courtesy of the Orlando Sentinel Communications Company.

production and related fields. More specifically, it is a text for *video graphic designers*.

Today's video graphic designer may be found preparing graphics for a news broadcast or designing the introductory titles for a situation comedy. He or she may be preparing artwork to be integrated into a computer graphics system as well. Business and industry are making better and more frequent use of video graphic designers in the preparation of color slides. Standard Oil in Chicago has pioneered in the use of computer graphics that include maps, charts, and graphs. At General Electric, an animation system called *Genigraphics* allows the staff to produce color graphics, interacting artwork creation with fully automated film recording under software control.

Several of the nation's largest newspapers are experimenting very successfully with the use of computerized graphic systems that allow for classified ad information and even full newspaper pages to appear on the screen.

Graphic design for the screen serves as the structural foundation of this book, with detailed analyses of the effects of space and distance among elements, the use of effective typography, color applications, the use of art and photography, the creation of promotional graphics, and the integration of graphics with today's technology.

Never before has the field of video graphics afforded greater possibilities for its practitioners.

Television graphic design is quickly developing as a discipline, one that requires a combination of skills: the ability

Frame G. ABC News Graphics staff at work: a conference for a rundown of newspieces in the first draft of the show.

Frame H. Graphics director going over graphic elements with designer.

to use space, type, art, and color—combined with sound—to convey a message in the fastest, simplest, and most effective manner. Television allows its practitioners to actually present coverage of an event live, but the demands and pressures of being creative while meeting a deadline are a reality of the business. This book emphasizes the challenge of creativity under pressure. Most important, it is our intention to inspire students of mass communications and design, as well as those presently working in the field, to consider possible careers in video graphic design.

References

1. Jeffrey Schrank, *Understanding Mass Media* (Skokie, IL: National Textbook Co., 1975).
2. Sandman, Rubin, and Sachsman, *Media: An Introductory Analysis of American Mass Communications* (Englewood Cliffs, NJ: Prentice-Hall, Inc., 1976).

VIDEO GRAPHIC DESIGN

THE IMPACT OF VIDEO GRAPHIC DESIGN

Few fields operate today without at least some computer-generated assistance. Household computers and word processing systems—most armed with a compatible television screen attachment—have steadily increased in popularity in the United States. Television graphics, if defined in its broadest sense to include any visual images projected onto a screen, become quite significant, especially in light of the technological advances of the past few years.

In fact, it is difficult to imagine how those pioneer video graphic artists managed to operate during the late 1940s and early 1950s. With very rustic tools (such as India ink, rubber cement, and Zipatone adhesive sheets, which were replaced by the currently used felt-tip pens), these artists were limited to placing drawn "Hollywood cards" in front of the camera in order to produce graphic art. It was, of course, a carryover from film, except that when displayed on a much smaller screen, the impact was weakened.

Those were the days before rub-on transferring letters, punch-out lettering, and, of course, electronic character generators. The only method used for creating type was applying photo type to an acetate sheet. There was little change in this technique until the mid- to late 1960s, when the art supply field began to expand significantly, accelerating to the level of sophistication that electronic devices have reached today. Those early attempts at graphic design for television laid the groundwork for the methods that video graphic designers use today.

British television designer Roy Laughton refers to the early period of television graphics as one of "sterility," attributing this

Frames 1–3. These illustrations from the so-called Hollywood era display plenty of imagination but also a lack of knowledge of how to use graphics in the new medium of television.

time to a lack of understanding of methods of producing effective television graphics. "Wallpaper backgrounds for titles were rich in texture but not in imagination. . . . Glitter flowed like water. The epitome of success was a roller caption filled with every conceivable style of typeface, laboriously hand-lettered and decorated with stars, spots and spangles from top to bottom."[1]

Of course, one element that helped the pioneering designer of television graphics tremendously was the novelty of the medium itself as perceived by the viewer. The mere presence of a picture on the tube sufficed to capture and maintain the attention of the audience. Today audiences are more demanding, especially because such a large segment of the viewers grew up exposed to an abundant dose of television.

Laughton's vivid description of those early attempts at providing television graphics is still applicable to some material that passes for "graphics" at some local television stations. Although technological advances make almost any conceivable graphic strategy possible, the training of video graphic designers still leaves much to be desired. Most colleges and universities in the United States do not include specific courses on the topic. Many practitioners confess that they are self-taught, having learned by doing; many come to the field of television graphics after having had successful careers in the print media. At some television stations graphic directors prefer to hire designers with a background in poster techniques, because the principles of boldness and simplicity are common to poster and television design. In both instances the goal is to achieve immediate impact through the use of visual elements.

There are four major areas in which a video graphic designer must be skilled:

1. design
2. the technology of video
3. typography
4. characteristics peculiar to the medium of television

These areas will be analyzed in detail throughout this book.

DISPLAYED MATERIAL

Video graphic designers are ultimately concerned with the image that is displayed on the screen. The spatial territory he or she must consider is a limited one: a screen area with clearly defined height and width characteristics. Like the architect who sketches, explores, and finally develops a design strategy on a blueprint, the video graphic designer begins with a sketch of a basic idea which is developed and executed with the aid of such devices as video input, photography, image enlargement and reduction, still-frame storage, character generators, and the effects possible through color coordination, echoing, and art.

In spite of this marvelous technology, the spatial limitations of the screen remain problematic. The aspect ratio of a classical television screen is in the landscape ratio of 3:4—that is, three units high by four units wide.

Experienced video graphic designers do not allow themselves to be discouraged by the limitation of this 3:4 aspect ratio. If a graphic must be used that is not in the specific 3:4 ratio, the designer may choose to zoom in closer with the camera and fill the screen with only one aspect of the graphic. Because the 3:4 ratio is approximately the same size as a 35mm slide, however, it makes the use of slides highly desirable if projected horizontally. Video graphic designers soon learn that they are not dealing with a 20-foot canvas but with a landscape design in which all elements must be contained within a certain size; the type cannot be too small and a sense of unity should prevail throughout.

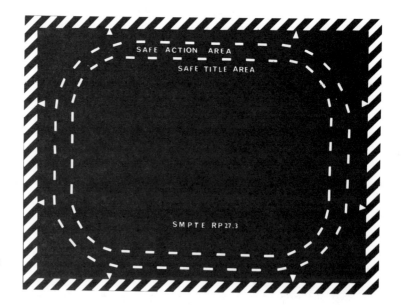

Frame 4. The aspect ratio indicates that all graphics must be designed in the proportion of three units high by four units wide.

Aspect ratio considerations for the television screen pose particular problems for the designer that do not arise, for example, in motion picture graphic design. However, generally most video graphic designers do enjoy more creative freedom than print media designers, who must concern themselves with line work, halftones, color ink, printing quality, and reproduction.

SOUND AND MOVEMENT

Effectively displaying an element on a screen is not the designer's only concern. He or she is also interested in determining the impact that a visual element will have, if it will convey the intended message, and, ultimately, how one element will relate to another and to *sound*.

Alan Wurtzel refers to television audio as the horse that trails behind the cart. "For years, producers and directors have devoted most of their attention to the visual portion of the show and given little consideration to the sound. . . . Regardless of how well-produced a show's video might be, if the audio production is weak, the overall quality of the show will suffer."[2]

Sound is vital to video graphic design, a characteristic that separates it from print design. The most effective use of *audio* is when it is accompanied by *movement*. Sound and movement, then, become the framework of the video graphic designer. No video graphic designer ever completes the design of a frame without considering the sound that will accompany the visual images.

There are strong perceptual links between *audio* and *video*. Both depend on *intensity* for their impact; whereas the intensity of a sound depends on the amount of energy used to produce pressure variations on the air, the intensity of a visual image on the screen depends on its overall effect on the viewer and on how well-unified the various elements within it appear. Even our reaction to loudness is based as much on a product of audio as of visual perception. We have all seen visual images that could be described as "loud" (many clashing colors, exaggerated type, distorted images). Although the term *pitch* usually refers to sound frequency (high-pitched sounds and low-pitched sounds), it can also relate to visual elements, from a typographic viewpoint. For example, bold and extended typefaces yield a high-pitched "sound," whereas thin and light typefaces convey the equivalent of a low-pitch sound.

Frame 5. This design could be described as "loud" because styles of type are mixed and the available space is crowded, creating a confusing effect.

Frame 6. The same material shown in the preceding frame but arranged in more of a "low-pitched" design.

SYMBOLIC THINKING

In addition to sound and movement, symbolic thinking is an important tool of the trade for the video graphic designer. Symbolic thinking refers to the use of visual images that convey the meaning of a message at a glance. In a medium that does not allow for any waste of time as it informs and entertains its audience, the speed with which the viewer captures the intended message is an important factor.

Symbolic thinking was first used by the primitive man who, in his desire to communicate, scribbled symbols on a cave wall or tablet. These hieroglyphics reveal a lot about ancient cultures. There are objects of symbolic thinking in our daily lives as well—for example, traffic signs, bottle labels, commercial or promotional logos. With more than 5000 languages and dialects in use today, it is no wonder that graphic designers constantly attempt to come up with what Alan Fletcher refers to in *Graphic Design: Visual Comparisons* as "an understandable symbology." As a symbolic thinker, the video graphic designer must conceptualize an event or message in a creative way. The designer who thinks symbolically avoids visual clichés and the mundane. For example, when school desegregation became a nightly topic on television news, many designers depicted the story by drawing an illustration of a group of children (an image that could be applicable to any number of stories involving kids). However, one designer chose the symbolic approach, conceptualizing the situation through the use of a schoolhouse painted half-black and half-white. This simple yet creative approach had a far more profound effect than a photograph of a classroom of children.

Similarly, during the Vietnam War, television screens were inundated with visuals of soldiers fighting in the field. One of the networks singled out an isolated map of Vietnam, an American flag, and a soldier. This graphic conveyed American involvement instantly. Consciously or subconsciously, it became a symbol for many viewers.

Many news events become recurring stories; and the responsibility of the effective video graphic designer is to create visual *images* that will also become memorable. For example, such newsworthy events as the Royal Wedding of Prince Charles and Princess Diana, the Pope's visit to Poland, the launchings of the *Columbia* shuttle, and the MX missile debate all became the subject of specialized graphic treatment.

The tools of symbolic interpretation for video graphic design are simple ones—in fact, the simpler the more effective as a communication strategy. Successful video graphic

Frame 7. Vietnam War graphics.

Frame 8. New Federalism graphics.

Frame 9. Symbolic thinking about the Atlanta child murders is expressed here through the use of simple elements, as executed in a Dubner CBG computer-assisted graphic.

Frame 10. This design effectively combines photographs of well-known faces, the British flag, and a typeface that fits in with the mood of the story.

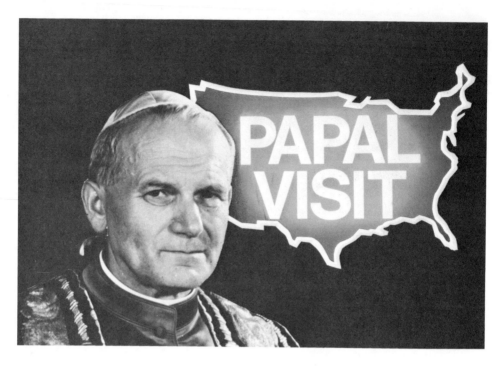

Frame 11. This design makes a simple but effective graphic statement.

Frame 12. Computer-assisted graphic for a story about the MX missile plan.

Frame 13. *Columbia* shuttle frame relates the image of the carrier, its mission, and a typeface that became easily recognizable to viewers.

Frames 14–23. These frames show action, movement, and texture.

Frame 14. This design shows visual movement and action.

Frame 15. Texture as a primary visual element.

Frame 16. Content prevails in this frame, which conveys symbolic representation through type.

Frame 17. The type selected here conveys the mood of the story at a glance.

Frame 18. The Cuban flag in the background communicates a headline easily.

Frame 19. Movement contributes to this design in order to tell a dramatic story.

Frame 20. Few viewers would need a verbal introduction to ascertain what this story is about.

Frame 21. Designers capitalize on the use of trademarks whenever possible.

Frame 22. The perspective of this illustration and the way it blends with the type create strong visual impact.

Frame 23. This frame includes movement, action, and symbolism—all of which are unified by a typeface that is easy to read and fits the tone of the story.

designers rely on the *content* of a message as their primary source of visual inspiration. Let the words give way to the visuals. Read the script carefully, paying particular attention to words that imply *movement* (up, down, sideways, parallel); *action* (conflict, clash, ultimatum, struggle); *color* (red tide, yellow journalism, black/white); and *texture* (fuzzy, velvety, rough, soft).

Promotional Graphics

Promotional graphics are probably the best examples of the impact symbolic thinking has as a video graphic design tool. It is through its promotional frames that a television station establishes a sense of identification with its viewers. Most logos aim at familiarizing people with a company, organization, or institution. Television stations are no exception; they emphasize a channel number, the geographic location of a station, program promotion, or simply a color.

Some stations vary their promotional presentation, keeping the same basic design but changing elements (type, color, size of illustration) in the design.

The illustrations presented throughout this chapter reflect the scope of creativity and the range of talents expressed by video graphic designers. These examples are also represen-

Frames 24–27. These frames show promotional logos for television news programs, station identification, and scheduled programs.

Frame 24. Various ABC News show titles.

Frame 25. Additional ABC News show titles.

Frame 26. Promotional graphic for an entertainment show.

Frame 27. Although not one of the most popular strategies, this Mondrian approach works well in this particular case, a sort of photographic montage that tells more than one story within one frame.

a.

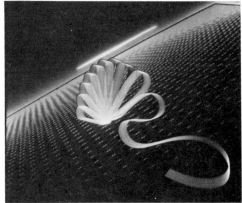

b.

c.

Frame 28a–d. These promotional frames, designed by Paul Fuentes, Charles Blake, and Elaine Zeitsoff of NBC-TV, New York, combine a familiar symbol with type for instant communication.

d.

VIDEO GRAPHIC DESIGN

Frames 29–32. These frames show the use of displayed logos. They are the work of designer Tom Mares, KOA-TV 4, who created the visual image of the station by designing the numeral 4 with a circle around it. This symbol is displayed on transportation vehicles, cameras, and stationery.

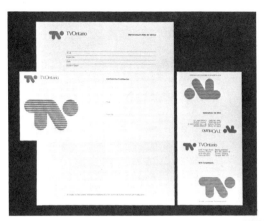
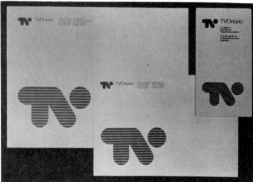

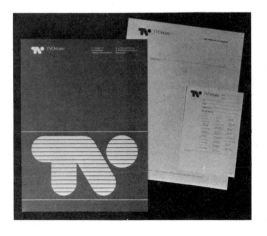

Frames 33–36. These frames illustrate how TV Ontario, of Toronto, carried a visual identity theme throughout their printed material: labels, tags, letterheads, envelopes, etc.

tative of the close cooperation that must exist between the designer and the director, producer, and other production personnel.

In *Television by Design,* Richard Levin reminds designers that their work accounts for 10 percent of all transmission time, a figure that is probably even greater today.[3] Pictorial elements exist to enhance content, to present a clearer picture of a message, to lure the viewer who may be watching but not listening, and to clarify meaning.

Video graphic designers accomplish this by combining effective video with synchronized audio, by conceptualizing their designs through symbolic thinking, and by emphasizing simplicity and boldness.

References

1. Roy Laughton, *TV Graphics* (London: UK Studio Vista; New York, NY: Van Nostrand Reinhold Co., 1966).
2. Alan Wurtzel, *Television Production* (New York, NY: McGraw-Hill, 1979).
3. Richard Levin, *Television by Design* (London: The Bodley Head, Ltd, 1961), p. 21.

DESIGNING FOR THE SCREEN

EFFECTIVE DESIGN

The following three frames represent the qualities of good video design: *simplicity, boldness, aesthetic value,* and *functionalism*. Well-designed frames also communicate a message at a glance, stimulate visual thinking, and convey the point of view of the message and/or program that they accompany. These qualities, common to every design medium, are derived from trial and error. Designers in the various media constantly experiment to achieve as many of these qualities as possible.

Simplicity

Long before man developed written languages he used signs and symbols to communicate. This, in a way, constitutes the first use of "visual communication." Each generation has enhanced such communication to the point where today most of what we learn reaches us through some form of visual communication; the best communicated visual messages are presented in uncomplicated and simple patterns. Even when there are several messages to be conveyed at a time—as, for example, a story involving two countries, one issue, and two different ideologies—the designer must strive for the simple graphic impact of *one idea* in order to create immediate graphic impact.

Graphic impact can be obtained by using different design tools—type, color, art, photography, or *combinations* of these elements—as long as one element dominates in order to create a single graphic impression. If the designer strives to achieve a look of simplicity, his work will be based on the *interrelationship* of elements, with visual ideas flowing but without exaggerated decoration and ornamentation. Posters, perhaps more than any other printed media, display the type

Frame 37. This frame conveys information through a map. The slanting of type creates the illusion of movement

Frame 38. This frame presents information via a graph/chart.

Frame 39. This frame presents a program title; only type is used, inspired by the familiar label of a wine bottle.

of simplicity that says it all at a glance. Such is the example that good video graphic designers follow.

Part of the necessary simplicity that accompanies the design involves reducing the graphic message to its simplest terms, while continuing to create a design interest in concept and execution.

Taking the complicated message and stating it in simple terms is the designer's challenge. For example, news stories such as those associated with the economy—which can be among the dullest when handled without creativity—should be presented in a manner that will keep the viewer *interested and informed*. Most elements within a story can be simplified and can get the point of the stories across through the use of instantly "recognized" graphic symbols: a bond, a passport, an IRS 1040 form, money, gold, and so on.

These symbols should look like the actual material in a quickly recognizable way, which can be accomplished by eliminating all unnecessary information—excessive lettering, detailed numbers, signatures, and so on—and preserving the *look*, not the fact, of the material stressed.

Boldness

Use of boldness in a design allows for immediate graphic impact. Boldness creates immediate clarity and recognition, and, when considering the obvious limitations of designing for television (such as the size of the screen and the time on the

Frame 40. A simple design by Jessie Tasencourt and John Ferlaine of WCAU-TV, Philadelphia. Film is combined with the number identifying the station; the design also conveys information about the film's programming.

Frame 41. The United States–U.S.S.R. nuclear arms race. The placement of the flags implies confrontation. The headline plays a secondary role to the graphic material.

Frames 42–43. These frames illustrate the design concept of using type as a secondary element; any other elements that may appear play a secondary role to type. This is also referred to as a "type attack."

Frame 44. A simple design created to convey a news message.

Frames 45–46. These are economy-related frames, with familiar trademarks and/or symbols displayed prominently.

Frame 47. Corporate ABC logo used on the air.

Frame 48. The logo for the ABC News program 20/20.

air), this quality becomes an even more important element to the designer.

An average television screen is approximately 12 by 16 inches. When viewed from a distance of 10 feet, that represents a mighty small cluster of visual elements with which to create an impression. Compare the screen to a magazine page, which, although measuring 20 percent smaller, may be viewed within a foot and the image (page) retained as long as the reader wishes to stare at it. The same may be applicable to a framed painting or poster. In contrast, the immediacy of a message shown on television must be conveyed in *bold terms* while creating a sense of *immediacy.*

The presence of boldness is probably perceived most strongly in the case of logos, where the recognition factor becomes the key to effective design. For example, in graphic

Frame 49. Logo for prime-time evening news show.

30 DESIGNING FOR THE SCREEN

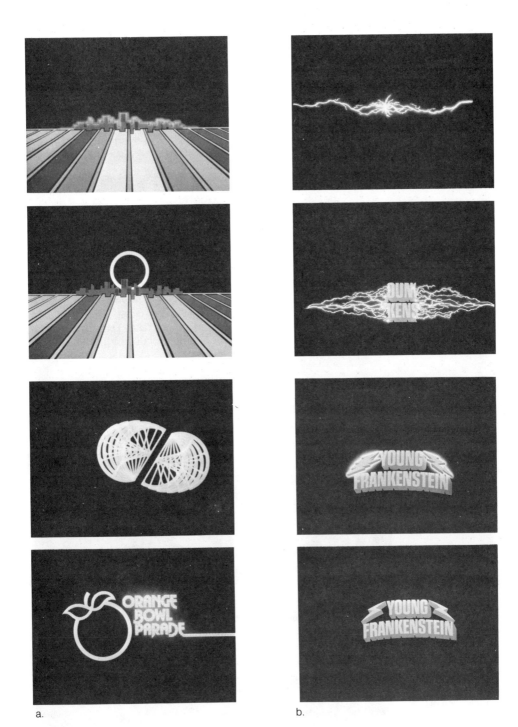

Frame 50a–b. These frames illustrate how the illusion of movement can be created through various designs.

design for promotion of programming, the element of repetition is used to emphasize a particular symbol that the station wishes to be associated with its image.

Aesthetic Value

In the final analysis, the success of any design—and especially that of a video graphic design—is dependent on its aesthetic value. An aesthetically pleasing frame combines the following characteristics:

1. *Perspective.* This shows the relationship of the various elements—that is, type, color, art and/or photography—as a *total* visual experience.
2. *Spatial distribution.* The *line of support* on the frame is where most elements come to rest; it may be either visible or invisible. The weight of the various elements should be distributed in order to prevent an "overload" on any one side of the frame.
3. *Clarity of movement.* Avoid creating more than one visible motion on the frame; don't take the viewer from circles to triangles to squares all in one visual take. The abruptness of such movement detracts from the clarity of the message to be presented.

Functionalism

This is where the most well-thought-out design strategy fails unless it adapts to the *medium* and the special *technological* circumstances in which it is to be used. *Functional design* implies adaptability of the frame to a particular situation, within

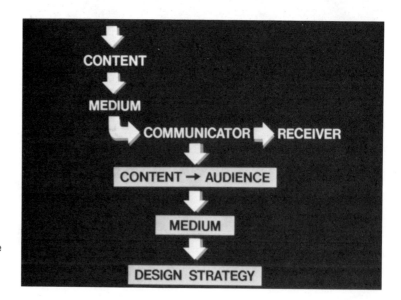

Frame 51. This is a special application of the communication theory chart, as applied to the designer.

other criteria as discussed in this chapter. A design too complicated to be carried out (one requiring equipment not readily available or one too contrived to convey the significance and clarity of a message at a glance) fails the functionalism test.

But functionalism extends beyond technological limitations into the area of *audience reception.* If we follow basic communication theory we discover the pattern of interaction between communicator (designer), medium (the screen), content (the message), and receiver (the audience). For our purposes let us take the traditional sequence of basic communication theory and adapt it to a *video graphic design situation.*

The revised chart shown in Frame 55 displays the content of a message and the audience as the first priority, setting the stage for the material to be designed. The medium (screen), which imposes some of the limitations described in this chapter, becomes the next concern and, ultimately, provides the perspective for the most appropriate design strategy.

Many experienced video graphic designers use this type of graphic thinking successfully, thereby reducing the content of a message to its very essence. For example, for a television mass audience a miner's helmet is a far more effective symbol than a miner; a thumbprint for crime is a better symbol than an entire illustration of a shadowed figure; displaying an American flag with a corner folded up, revealing a hammer and sickle, is a better way to convey a spy theme than using an elaborate illustration of a spy; a fireman's helmet is a more effective visual image than a shadow drawing of a building in flames; and a license plate with the word "rationing" on it sums up fuel shortages better than a line of cars at a gas pump.

PERSPECTIVE

The following two illustrations demonstrate the importance of perspective within the design of a frame; they vary in the degree to which they organize material as a visual unit. Notice that the same elements are used in both illustrations, but in the first the organization of these elements distorts the overall visual impact of the frame; it has too many "ragged" angles, which prevent it from appearing unified. The design shown in the next one establishes a sense of visual support whereby the elements are aligned on the left; the symbolic image of wheat is shown more simply here than in the other photograph.

Frame 52.

Frame 53.

Unifying the various elements within a frame gives a sense of "space" around the illustration. Notice how these two frames dealing with inflation create different impressions. The first one does not convey order. Three main elements (type, map and arrow, and dollar bill) are not linked graphically. In contrast, the next frame demonstrates what unification can do to make a visual statement faster and in a less complicated manner.

SPATIAL DISTRIBUTION

Frame 54.

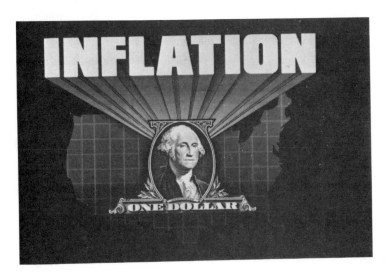

Frame 55.

DESIGNING FOR THE SCREEN

The two frames that follow demonstrate the importance of the spatial relationship of type within the headline in making the meaning of graphics clear. Why separate a headline into two areas of the space available, when positioning it flush right in one area can communicate more quickly. The more sophisticated approach to the art in Frame 61 makes it more effective.

Frame 56.

Frame 57.

The examples that follow show how a single statement, if presented in a clean, concise manner, can have greater impact and provide more clarity of meaning than the more complicated arrangements of various themes that vie for attention and eventually detract from fast, effective communication.

CLARITY OF MEANING

Frame 58.

Frame 59.

Frame 60.

Frame 61.

Frame 62.

Frame 63.

Designing for the screen involves *composition, layout,* and *graphic thinking* in the same way as for the print media but with some differences, mostly the result of those characteristics that are peculiar to the medium of television.

Experienced designers approach design from the following perspective:

1. Visualizing a design (sketching)
2. Experimenting with ideas such as enlarging, distorting, abstracting, integrating, separating, and adding
3. Executing a final idea

THE DESIGNER AND HIS CRAFT

All graphic thinking usually begins with some type of visualization—putting basic ideas on paper. "Show me, don't tell me," a design professor constantly reminds his students who insist on "discussing" their graphic ideas. Even the most tentative sketch will lead to more concrete ideas than would a conversation about a design. It is important for video graphic designers to practice free-hand drawing and to get in the habit of always carrying a sketch pad. Designers make this visualization step the starting point for the emergence of their ideas.

Visualizing

Sketching begins the creative process. It provides a foundation, a visual perspective from which to continue experimenting toward a final design. If the sketches drawn during the visualization stage are done well, complete with straight lines, defined spaces, texture, and other details—such as color and range of tonal quality—then the process of experimenting with those basic ideas is made easier and more productive.

This process of experimentation gives the designer an opportunity to solve problems as well, offering alternatives, new insights, and alternative choices. An early sketch may

Experimenting with Ideas

Frame 64. A hand-drawn sketch.

Frame 65. The final product from a hand-drawn sketch.

Frame 66. Miniature rough sketches help the designer conceptualize ideas, to be expanded upon later and turned into final frames.

immediately show that the design is cluttered, or unrealistic in terms of space availability. The value of a sketch is to provide visual evidence that the design is feasible, and that it adapts to the needs of the message while conveying an aesthetic and functional look. For designers who are not artists, the simplicity of representative motifs (such as circles to indicate a person, squares to represent a block of type, or shapes to indicate maps or graphs) should suffice during the early stages of conceptual representation.

One of the most important reasons to sketch and to experiment through simple pencil sketches is to get a visual perspective of the total package, considering that the sum of the various elements within the design—their relationship—provides the overall communication of the message quickly and clearly. Many changes will be made before a sketch is finalized and readied for final execution. The process of designing is neither simple nor short. It was mentioned earlier that the reason for experimenting through drawing sketches is to create a visual perspective. Such a perspective can refine an idea—improve it and polish it—by using several techniques, all of which should become familiar to the designer.

Enlarging After drawing the first frame opposite, the designer realizes that there is no contrast among the many elements within the frame. So she decides to enlarge one of the elements. The result is a clearly perceived center of visual attraction: one dominant element, with all other elements rendering support (see following frame).

Frame 67.

Frame 68.

Distorting Distortion has always been a tool popular with designers. An optical illusion can be created, much like the type that children enjoy—if only ephemerally—when they pose in front of a mirror that distorts the shape of their heads or length of their legs. When used appropriately, distortions dramatically alter our view of reality and hence bring about visual interest. The television screen is an ideal medium for using visual distortion, and the new technological devices currently available make the technique easier to achieve and make its reception more dramatic and dynamic.

Distortion can enhance the commonplace. A dollar bill can be dramatically elongated out of proportion; a tiny world spot, such as El Salvador, can become "monumental"—thereby adding to the symbolism of the story.

Frame 69. Close-up version of CIA logo, "magnified" to make news point.

Frame 70. Close-up (distorted) view of American flag.

Frame 71. U.S. flag made into arched arrow.

Frame 72. A distorted version of a map.

Frame 73. A distorted map and a close-up of an area.

Frame 74. A map of China and photos of party members juxtaposed.

Integrating It takes a designer with special talents to establish visual analogies—to incorporate elements that give the story a cohesive vision; for example, men and a map of their country. Integrating also means offering comparisons of characteristics; for example, a frame that shows unemployment up in the east and slightly down in the southwest.

Separating It is as important to integrate elements as it is to separate those elements that, although part of the same message, may not work well together. At this point, the designer must understand the nature of the message and its substance before making a final choice about which graphic elements he will use. Let's assume that the story to be illustrated deals with *solar energy*. If the story highlights the potential *savings* for the consumer, then the graphics should deal with monetary remuneration, as shown in the next frame.

If, however, the story emphasizes the *comfort* of enjoying a warm home environment even on the coldest day of the year, then the approach could be more personal, the point of view more relaxed, as shown in this frame.

Adding It is almost always easier to eliminate elements than it is to add them to a design. The reason for this may be based on the way we think in terms of our total visual experience. For example, how many times have you enhanced your personal look by eliminating a tie that didn't complement the rest of the

Frame 75. A dollar is a familiar symbol; however, the center of it can be redesigned to include a visual element pertinent to the content of the story, such as an image of the sun in the case of a story on the monetary advantages of solar energy.

Frame 76. An illustration for a story about solar energy that emphasizes comfort. The two ideas presented in this frame and in the preceding one would fail if they were presented together. The designer must study the available content—the "angle" of the copy—in order to arrive at a workable and meaningful design.

suit, or by removing an extra necklace that detracted from other jewelry? Eliminating is easier than adding. The same person who has eliminated his tie from a particular outfit will at other times stare at himself in the mirror, trying to figure out what's missing for that total look.

Eliminating often helps to unclutter a design, whereas adding presents the danger of creating just the opposite effect. However, many situations call for *additions,* and in some instances it is best to delete elements within the design.

Execution of an Idea The visualization and experimentation stages are the prelude to the final execution of an idea. All the work involved in sketching, enhancing, rejecting, and polishing pays off when the finished product emerges. Even if the viewers never realize the amount of work that goes into the conceptualizing and executing of each frame, the designer will enjoy the reward of his or her work in enhancing and attracting attention to what may otherwise have been another lost message. That, in essence, should be the goal of those persons who design for the screen.

TYPOGRAPHY FOR THE SCREEN

To many people the mention of the word "typography" conjures images of printed pages—the cover of *Vogue* magazine, or a graphic calendar inspired by the Bauhaus, and, of course, a front page from *The New York Times*. Typography is strongly linked to the print media, from the time of Gutenberg and his impact through the further evolvement of type through such renowned typographers as Garamond, Baskerville, Bodoni, and Goudy. In spite of the more traditional use of type in connection with the printed page, some information has appeared in type on the screen, beginning with silent films, which relied on type for clarifying meaning.

Today the level of typographic sophistication and choices of typefaces for the screen equals and even surpasses its use on the printed page. During the past twenty years television has progressed from menu boards (handpainted opaque title cards) to hot press to mechanical digital units to primitive alphanumerical displays to the present alphanumerical devices, which are as revolutionary as the invention of movable type and the printing press. As far as the video designer is concerned, typographic history is in the making, with no end in sight for that which may be achieved in the future as technology expands.

Never before has typography offered such flexibility and variety: The new alphanumerics (electronic typing machines that can supply on screen a variety of typefaces in multiple sizes and colors) offer the designer any type, in any color, with or without lines, upright or italicized, overlapped or normally spaced. It is now possible to have hundreds of type styles available for specific uses, all with the precision to cut through the video with *detail* and *clarity*.

Frame 77. Type produced on an alphanumeric machine.

Frame 78. Close-up of above frame showing the "jaggies" or broken lines caused by diagonals from electronic screen lines.

PURPOSE OF TYPE

The ultimate purpose of type is to make reading easy. Unlike a printed page, the television screen does not always require *type* as an element to make meanings clear; other visual elements and sound become more important in communicating information. But type can often help make the material on the screen more meaningful. Here are some commonly used strategies for using type on the screen:

1. To *enhance* the information presented visually. For example, quotes or excerpts from printed documents may accentuate what is presented, as shown in the next four frames.

"WE WILL DO ANYTHING TO KEEP THIS COMPANY ALIVE."

Frame 79. A quote in caps positioned at the bottom of the screen.

"...I had just woken up and it was dark and I think it was three men...right..."

Frame 80. A quote in caps and lower case positioned at the bottom of the screen.

TYPOGRAPHY FOR THE SCREEN

> "...I had just woken up and it was dark and I think it was three men...right...

Frame 81. A quote in italics.

Frame 82. The date is displayed in the upper left-hand corner of the screen.

Aug 12
<u>1981</u>

Frames 83–84. Some rules for type placement and usage at ABC News.

2. To *direct the reader* to a certain area of the screen, and to identify the content or categorize coverage, as indicated in the following two frames.

Frame 85. Location and title identifiers used at ABC News at the beginning of a story.

Frame 86. The words "Special Report" displayed as a slash.

TYPOGRAPHY FOR THE SCREEN   53

3. To *identify* a person on the screen, such as a news source or a reporter.

Frame 87. The name of a news source.

Frame 88. The name of a reporter.

4. To *place* the viewer in the geographic location where a report is taking place, as shown in Frames 96 and 97.

Frames 89–90. *Right* and *below,* examples of type used to identify locations.

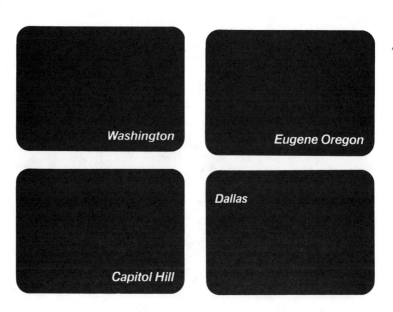

The type fonts shown in the frame below are only a handful of the fonts available to designers. Each font represents an acceptable, readable typeface; in each case the individual letters are easy to read, bold, elegant, and, most importantly, tested to work well on the screen. But it is at this point that the comparisons stop, because each typeface offers the designer a different set of *type values;* that is, visual appeal, appropriateness to content, and adaptability to other elements appearing simultaneously on the screen.

CRITERIA FOR SELECTING TYPE

Frame 91. A variety of type samples.

The competent designer selects a typeface on the basis of how these values affect the individual situation. Generally, good screen typefaces possess the following qualities:

1. *Readability:* Type that appears clearly on the screen, can be read at a glance, and offers good resolution.
2. *Impact:* Type that stands out regardless of what other elements surround it.
3. *Boldness:* Type that consists of bold, well-contoured letters reads better than light, thin characters.
4. *Simplicity:* It is a universal typographic rule that simplicity of style lends itself well to readability, but the statement takes on even more significance in reference to selection of type for the screen.

Let's examine the typefaces presented earlier in this chapter in terms of what they offer and how they can best be used.

Frame 92. *American Typewriter:* This typeface resembles typewritten letters. Each character is well-contoured and is most effective in lower case. It is often used when a story is presented as though the reporter were typing the story or event on his or her typewriter.

```
abcdeefghijklmnopqrstuvwxyz
ABCDEFGHIJKLMNOPQRRSTUVWXYZ
1234567890 (&.,:;!?'""-—-*$¢%/£)
```

Frame 93. *Avant-Garde:* This typeface offers boldness, well-rounded letters, very open capital letters, and distinctive numerals.

```
abcdefghijklmnopqrstuvwxyz
ABCDEFGHIJKLMNOPQRSTUVWXYZ
1234567890(&.,:;!?""""-—-·()˙$¢%/£)
```

Frame 94. *City Compact Bold:* This is a serif typeface; it is condensed, displays a solid look, and conveys a feeling of stability and seriousness.

```
abcdefghijklmnopqrstuvwxyz
ABCDEFGHIJKLMNOPQRSTUVWXYZ
1234567890 [&.,:;!?'''-·$c%/]
```

Frame 95. *Windsor:* This typeface creates a classic, elegant look.

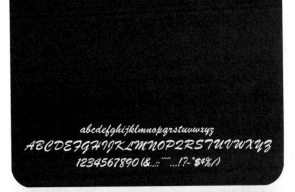

Frame 96. *A Script Face:* Script faces do not read well on the screen—the individual characters lack boldness and impact, and often fade into the background. The exception—and an attractive one—is using a person's signature to accentuate commentaries in the case of well-known personalities.

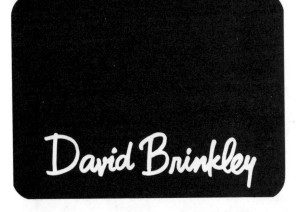

Frame 97. *Left,* example of David Brinkley's signature.

Frame 98. *Below left,* An example of an ornate face. Extremely ornamented faces create too much visual tension on the screen, compete with other elements on the screen, and fail the readability test.

PLACEMENT OF TYPE

The television screen is seldom totally dominated by type. In most cases typographic elements appear as secondary elements, serving in a supporting role to highlight graphic images. The aspect ratio of the screen (see Chapter 1) limits placement of type as well. It is vitally important to position type within the essential area of the screen. Television designers place type according to how the type relates to other elements appearing on the screen. Some examples of placement alternatives follow.

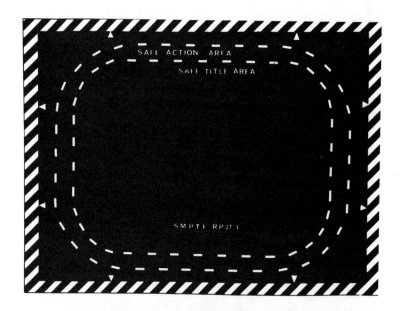

Frame 99. Aspect ratio (essential area).

Frame 100. Placement at the bottom of the screen.

Frame 101. Placement at the upper left of the screen.

Frame 102. Placement at the top center of the screen.

Frame 103. Placement at the center (middle of the screen).

Frame 104. Placement at the top and bottom of the screen.

Frame 105. Slashed on the upper left of the screen.

CREATIVITY AND TYPE

The typographic tools presented in this chapter so far are limited only by the creative inventiveness of the designer. With the help of technological advances that provide good screen resolution—even for the most delicate of typefaces, and allow type to rotate, positioning it at whatever angle the designer desires—there is no limit to what a creative designer can achieve with type.

Let's examine some of the possibilities of typography-dominated graphics.

Color

Using color contrasts is an effective way to distinguish between lines of type. Yellow cuts through most video quite well, and the use of a black drop shadow on all lettering separates the type from the video background.

Photography and Illustration

Photography and illustration may be used effectively, especially when combined with typographic elements. A silhouetted full-colored object combined with graphics provides a full range of graphic contrast. Both the photograph and the type complement each other, as shown in the next four frames.

Frame 106. This is an example of a show logo which was created as a font symbol so that it could be retrieved with one keyboard punch.

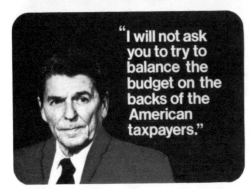

Frame 107. A combination of a photograph and a quotation.

Frame 108. This frame combines type and photography to convey a dramatic mood.

Frame 109. A photograph with type on a slash.

Frame 110. A map and type combination.

Frame 111. A court drawing, with type superimposed.

The dominance of typography—when used creatively—can be as powerful and dramatic as any other graphic device. With the help of videographics—a microcomputer-based color TV graphics generator which combines a character generator with a real-time animation machine—enables the designer to create individual frames and to play them back, creating the illusion of movement. In addition to the more than 200 type fonts available in the computer, the operator is able to create any new font and place it into the computer's memory. Not only can characters be popped onto the screen, but words can be written across the screen as though written by hand.

Special effects

Frame 112. *Liberace piano:* A graphic piano provided an introduction for a story about Liberace. Its keys appeared to move, an effect created by depressing them, highlighting some on a screen, darkening others in frames, followed by editing.

a.

b.

c.

d.

Frame 113a–d. Charles Blake's design of the *Fame* logo (NBC-TV New York) involves movement to reflect the nature of the program. This is an example of typography working to establish a mood.

Frame 114a–c. The syringe graphic series shows droplets changing into letters which appeared on the television news show *20/20*, spelling the title of a story.

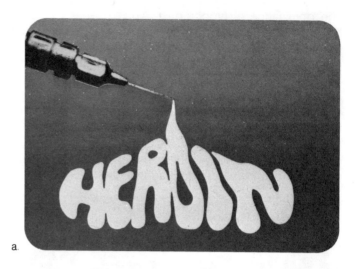

a.

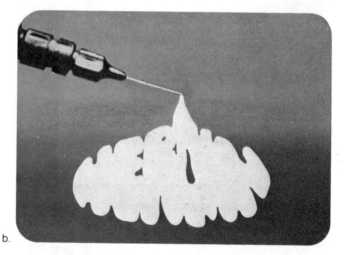

b.

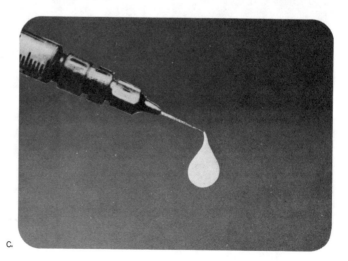

c.

The role of typography assumes special importance in the case of certain news events. Here typography is used to clarify meaning in such events as elections, economic news, the launching of space shuttles, and sporting events.

In most instances, these events require numbers, special symbols, and type to aid other visual elements on the screen. In such cases the designer must select typefaces that *assist* the other visual elements without creating an overpowering effect.

The following frames display some examples of designs that assisted in ABC News's coverage of contemporary events.

TYPOGRAPHY AND SPECIAL EVENTS

Frame 115. *Zoom logo:* The words "ABC NEWS" zoom toward the viewer from back to front.

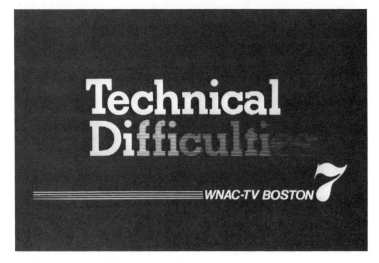

Frame 116. The message is clearly stated through the use of typography in these frames designed by Mary Lawrie and Robin White, WNAC-TV, Boston.

Frames 117–118. An effective design for election return figures.

Frame 119.

Frame 120.

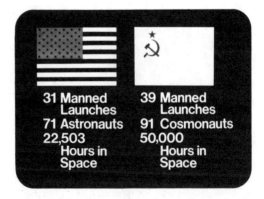

Frames 121–122. The launching of Columbia Shuttle, 1982.

Frame 123.

Frame 124.

Frame 125.

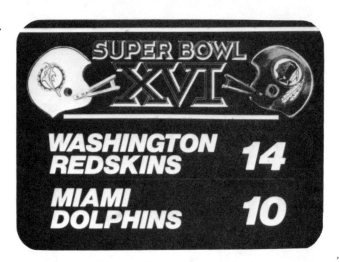

Frame 126.

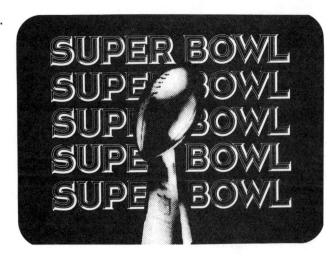

The alphabet's twenty-six characters are not the only typographic tools available to the designer. In addition to the alphabet, there are numerals, punctuation marks, arrows, squares, pointers, and stars, to name a few of these tools.

TYPOGRAPHIC SYMBOLS

Frames 127–128. Use of arrows.

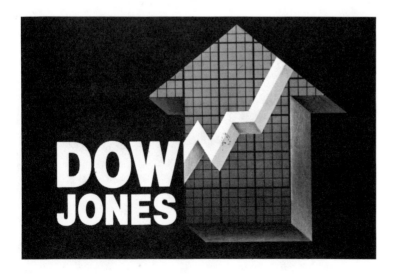

TYPOGRAPHY FOR THE SCREEN

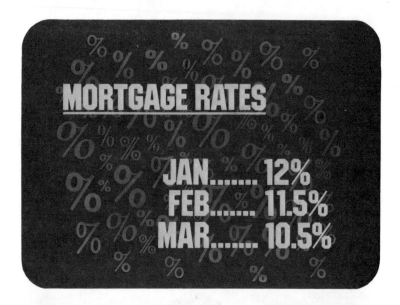

Frame 129. Use of percent sign.

Frame 130. Type has been used effectively to incorporate the essence of a story through the simplicity of a label title.

Frame 131.

PHOTOGRAPHIC DESIGN FOR TELEVISION

It is somewhat ironic that television, a medium so saturated with essentially photographic images, continues to capitalize on still photography to convey information. And although it's true that still photographs do not monopolize the screen, when used effectively a dynamic photograph can enhance the design and scope of the message presented.

Television represents the electronic display of moving pictures through film, tape, and in live form. It is difficult to compare the somewhat limited visual impact afforded by a still photograph to moving images on the screen. But in fact, the average person who reads newspapers and watches television news has an opportunity to make the comparison on a daily basis. For example, a person may read in the morning newspaper about the visit of an important foreign dignitary to Washington, and sees a photo of this person shaking hands with the President. The evening television news will "extend" the reader's visual experience of the event by adding the qualities of *motion* and *color* to the scene. Suddenly, the still photo—that frozen image of a time frame—moves; hands actually perform a handshake, feet move, and reality as we know it is no longer the stationary image of the still photograph in the newspaper.

Why, then, should television graphic designers be concerned about using still photographs?

Still photography plays a very important role in television graphic design, especially in the illustration of news events. Many news stories require a still photograph; in many instances tape or film are not available but still photographs are.

The selection of a still photograph for the screen becomes the designer's task. It is not an easy choice to make, because the photograph selected must be of dynamic quality, and its on-the-screen treatment must be enhanced for maximum effect.

CRITERIA FOR SELECTING A STILL PHOTOGRAPH FOR TELEVISION

A Sense of Urgency Many news events happen unexpectedly, and consequently there is no opportunity for television camera crews to be on hand to record them. However, because photography is a popular hobby of many people, it is more likely that an amateur photographer rather than a cameraman will be present at the scene of a news event. This is the reason why news photos often become the only available visual record of important events. When John Hinckley, Jr., shot President Reagan and four other persons, including presidential press secretary Jim Brady, a photo of Reagan displaying an expression of pain was used, along with footage, to convey the mood of the shooting scene. The designer could have made other choices—such as using a symbol of the U.S. President, a head shot of the President, or a gun—but nothing could equal the news photo which showed the President seconds after the bullet hit him.

A Sense of Realism Sometimes nothing can provide the sense of realism and visual authenticity as well as a photograph. For example, in telling the story of a New York City garbage strike, no other visual strategy could say as much graphically as a color still photograph of many stacked garbage cans and plastic bags. This conveys a feeling of realism that art cannot communicate effectively.

A Sense of Feeling Sometimes the nature of a news story's content lends itself to the type of photography that creates a feeling or mood instantly for the viewer. For example, a documentary-type essay exposing the depth and breadth of America can be beautifully illustrated through photographs of scenes such as the Golden Gate Bridge, the Statue of Liberty, and wheat fields in the Midwest.

Frame 132. This photograph was used to illustrate a garbage strike in New York City.

Frame 133. Both the type used for the word "Elderly" and the photo combine effectively to create the feeling that best expresses the content of the story.

For more direct news stories, such as political elections, a good photograph can capture the feeling and spirit of a hand raised in victory or the tired face of a defeated candidate.

And what better way of presenting the mystery of a volcanic eruption, such as Mount St. Helens than with a still photograph to accompany the headline of that story.

Frame 134. A photograph of Mount St. Helens that gives a dramatic perspective.

STRATEGIES WITH STILL PHOTOGRAPHS

Different news situations lend themselves to individualized strategies aimed at highlighting and maximizing the use of each photo in direct relation to the content it is supposed to enhance.

Silhouettes

Silhouetting—dropping the background of a photograph—produces one of the most effective treatments for head shots as well as for objects. The silhouetted full-colored object in conjunction with other graphic elements, such as type or art, produces a rich, full-range graphic on the screen. Silhouettes are functional in that they force the designer to eliminate from the photograph such extraneous material as backdrops, crowds, microphones, and other distracting elements. The designer must take a careful look at a photo before deciding to silhouette it. Some photos may be better left as full frames than as silhouettes, especially if two or more subjects are photographed. When silhouetting head shots, a dark (or black) background projects the head out to the best advantage.

Silhouettes are not limited to head shots. Such symbolic objects as the sun, the moon, snow, a jail door, a fireman's helmet, a shopping cart with groceries, a cash register, a church, and vehicles can come in handy on deadline.

Frame 135. A combination of a silhouetted head shot and a related map gives the viewer an instant idea of the elements of the story.

Frame 136. A silhouette of a photo enhances the design.

Frame 137. A silhouette of a map.

Cropping How a photograph is cropped determines the impact it will have on the viewer. Cropping techniques for video graphic design are basically identical to those used in publication design: eliminate nonessentials, enhance the main person and/or object of the photograph that tells the story, emphasize motion, bring attention to detail, and avoid too many images per frame.

Regarding head shots, the general practice is to allow a small amount of space above the head and down to just below the knot in the tie. Tight cropping of a well-known face—a practice referred to as "mask" cropping—can often give dramatic emphasis to a story; unfortunately, not many television designers make use of this strategy.

For head shots of well-known personalities, high contrast in color photographs does not "read" too well on the air. The designer must make sure that the lighting range is not extreme and that color saturation is not too heavy, especially toward red tones, which tend to appear blurred on the screen.

Frame 138. A full screen head shot.

Frame 139. A head shot positioned above the shoulder of the commentator.

Because content should dictate photo selection, it is important that designers secure a wide range of head shots of well-known personalities—ones likely to recur in the news—to allow for adequate facial expressions with the changing news events. The most *neutral* facial expression is usually the safest choice; however, for dramatic impact the *angry,* the *smiling,* or the *unhappy* portrait, as appropriate to the news story, can make the point of the story more directly.

In general, head shots should be well lighted, cropping should be tight, and as much of the screen should be used as possible. If the photo is boxed and used over the commentator's shoulder, the head can be cropped to leave as little space as possible around the photograph. The name of the person in the photo can be on a color bar below the photo or to the left or right of the head.

Integrating Photos and Other Visual Elements

The interplay of full-color or black-and-white photos with line and air brush work can be visually appealing. Art and photography complement each other, and the combination of a good photo with, perhaps, type or even computer-generated art can offer a very sparkling graphic display to enhance any content.

Frame 140. A story on the British economy.

Frame 141. Photo/type combination.

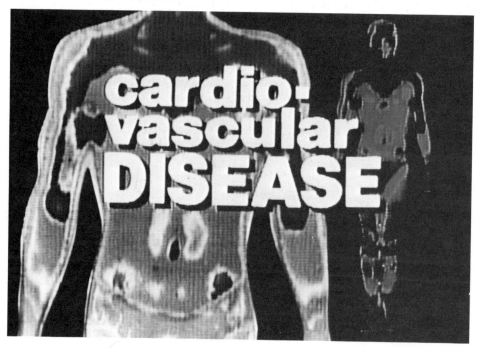

Table-Top Photography

Table-top still life photography can add a new perspective to television graphic design. Shooting an object against a flat background with lighting to give the dimension required for a specific design, and adding type or other elements later on the recorded image, can offer a richness and freshness that a drawing of the same object might not have.

Line Conversions

Line conversions (the elimination of all halftone quality from a photo, leaving only black-and-white areas and lines) provide high-contrast blacks and whites, and can also give a unique arresting look when incorporated with other art. Action sports, head shots, and common symbolic objects all can be offered a different perspective on the screen through line-conversion treatment.

DOCTORING PHOTOGRAPHS FOR THE SCREEN

The techniques of touching up photographs for the screen are similar to techniques used for printed pages. The treatment a photograph is given can make the difference in its impact: A photo can be cropped dramatically, unnecessary elements removed, and an important element highlighted or realigned. It takes a designer with an eye for detail to make the best and most exhaustive use of each photo.

    The following techniques—not all of which are necessarily used simultaneously—guarantee that each photo will receive attention to exploit its visual potential.

Shoulder Adjustment

This technique enhances the appearance of a person in a photograph. Compare the next two frames: The first is an original in which the person's shoulders seem to overcome his head; the second is an adjusted version.

Frame 142.

Frame 143.

Retouching Nothing can be more detracting in a photograph than the presence of such unwelcome elements as microphones sticking into a person's chin, or an arm protruding from or into the photo. The designer can retouch to eliminate such distractions. Compare an original photograph with a retouched photograph (following two frames).

Frame 144. Unretouched.

Frame 145. Retouched.

The realigning of a photograph involves "movement" of elements to display them within a proper perspective of distance. For example, when two astronauts—one short, the other one very tall—stand side by side, a simple realignment can make the two persons appear to be closer in size. Compare an original photograph with a realigned photograph version (the next two frames).

Realignment

Frame 146.

Frame 147.

PHOTOGRAPHIC DESIGN FOR TELEVISION

Frame 148. Close-up of a hand and test tube for medical reports.

Frame 149. A police badge.

Frame 150. For construction-related stories: a tight shot of bricks and a trowel.

Sizing

If there is any question about how to size a photograph, the designer should move in as tightly as possible without losing the essential information provided by the photograph. In sizing a photograph for the screen, one must take into consideration the fact that "type" almost always accompanies the photo, and space must be allocated for it.

DEVELOPING AND MAINTAINING PHOTOGRAPHIC FILES

An extensive and easy-to-use file of photographs is one of the graphics department's most important resources. The very nature of news, the speed with which it appears, and the limited time available to locate suitable materials to display it create a need for a well-stocked and easily accessible file of photographs.

One of the easiest ways to start such a file is by creating photographic graphics based on generic elements. Consider, for example, the following subjects:

Even though today many photographic file photos are being replaced by the frozen frame from tape and retrieved and pretaped from the event of the day (which provides accuracy of appearance to the news item), many graphic arts departments in smaller, local television stations continue to extract their photographic sources from the traditional places: Associated Press, United Press, Image Bank in New York City (which provides personalities and places), Black Star (a similar service), Magnum Photos, and Gamma–Liaison.

Every television station should have handy books with historical illustrations for stories that deal with the past. There are many fine books in this category, including some devoted to regional or local areas of the country.

The trend to still photographs from tape is increasing, and with the ability to adjust size and position on the screen, it is possible that the photographic file of the not-so-distant future will look more like a videotape library. But it is unlikely that the still photograph as we know it today will completely lose its value in television design.

ART AND ILLUSTRATION FOR THE SCREEN

The use of art and illustration in television has increased rapidly and dramatically during the past ten years. Sometimes art and illustration contribute to clarify a news event, as in the case of a sketch to indicate the scene and sequence of a bomb explosion or the path of a tornado. Often the illustration assumes the role of a poster, creating immediate appeal for a story. Regardless of its function, an illustration designed for the screen should possess the following characteristics:

    It must have immediate visual impact
    It should clarify and identify the essentials of a story
    It should be distinctive
    It should be emphatic

Frame 151. A sketch of a scene in a courtroom.

Frame 152. A sketch of a scene in a courtroom.

WHEN SHOULD ILLUSTRATIONS BE USED? Certain news events call for the use of illustration rather than photography. Cameras are not permitted in courtrooms, but artists are. Some of the best use of art and illustration seen on the screen is the result of courtroom sketching.

The important criteria in courtroom art are the abilities of the artist to capture likenesses and to sketch rapidly. The artist eventually returns to the studio to add color and detail to the sketch. It's also important for the artist to capture the architecture of the courtroom, in addition to the leading characters involved in the trial.

Certain events not captured by photographers can be "rescued" graphically by the artist, who attempts to reconstruct for the viewer what and how something happened. Such was the case when John Lennon was shot; the artist captured the event for the viewer in this dramatic sketch.

Frame 153. A sketch of John Lennon's assassination.

Frame 154. This sketch offered viewers a glimpse of the Pope the moment he was shot.

Frame 155. The Falkland Islands war of 1982, with its meager photo output, was the type of event that was best covered visually through illustration.

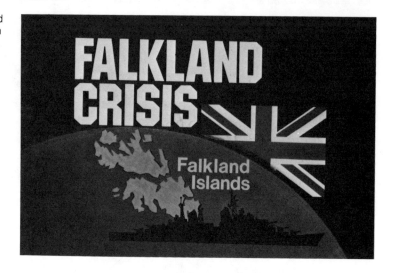

Complicated data can be clarified easily for the viewer through a well-organized frame composed of color, type, and illustration. The same applies to scientific and medical information, for which the artist creates a more meaningful frame by taking the information, abstracting it, and presenting the highlights, as shown in Frame 164.

Frame 156. An illustration of technical data.

Remote events, too, lend themselves to the use of art rather than photographs. Viewers learned about the assassination of the Egyptian president Anwar Sadat with the aid of sketches. You can sense the impact of this dramatic event as perceived by the artist in the frame below.

Frame 157. Sadat's assassination.

Caricatures have always been a popular form of illustration. Throughout history newspapers have delighted readers by featuring distorted and exaggerated visions of famous persons. Magazines, too, have capitalized on this art form. The use of caricatures on the screen can be just as effective; the application of personal sketches is varied, ranging from newscasts to talk shows to any type of visual presentation.

Sometimes a biographical presentation makes use of drawings on the screen. At ABC News this technique became a way of taking the reader to Pope John Paul's childhood days in Poland, or to Premier Menachem Begin's early days in Israel.

In fact, it is in the type of situation where personalities are involved that art and illustration not only can enhance screen presentation but also can provide much-needed contrast. In a show full of film and photos the introduction of a person's "sketch" can provide the viewer with relief from the overabundant use of other visual images.

Frame 158. This sketch was included in the biography of Israel's Menachem Begin.

Frame 159. Begin's life illustrated through sketch.

Frame 160. This illustration adapts to the familiar symbol of the American flag and map, and connotes the image of "fat" even before the headline tells the viewer about the content of the story.

As mentioned earlier, the visual presentation on the screen complements the audio portion; and illustrations can be a very effective "liaison" between the two. For example, in the case of news coverage that attempts to reconstruct an air disaster, recordings of pilot-to-tower conversations depicting the last few seconds of an ill-fated flight can be augmented by an illustration in order to help the viewer conceptualize what was happening while the conversation took place.

When Not to Use Art Some news items are much more appropriately presented as photographic events (see Chapter 4) rather than with art and illustration.

Such recent events as the birth of a son to the Prince and Princess of Wales required a photographic display. An art rendition would have given authenticity to the event. Similarly, the wedding of Prince Charles to Lady Diana Spencer certainly had to be captured photographically.

What Media to Use for Screen Art Different artists and art departments have unique preferences for executing screen art. The particular medium used is of secondary importance to the impact the screen art makes in conveying a visual message instantly. Whatever medium the artist works best and fastest with, and feels most confident working with, is the medium that should be used.

Some artists prefer to use pastels; others prefer pencils, with magic markers for color; some artists even prefer to use water colors or colored pencils. All of these are acceptable media, each providing the artist with a distinctive, unique richness of style. In fact, a combination of these various approaches often creates a variety of presentation on the screen.

THE DESIGN AND USE OF MAPS AS INFORMATIONAL GRAPHICS

Maps bring the viewer closer to the geographical specifics of a story, providing an opportunity to expand the viewer's perspective of an event in terms of its *location,* the *surrounding areas,* and *how it relates to the viewer's situation.*

Maps have always held a special fascination for people. A person's first encounter with maps is in an early elementary-school textbook. Maps also appear in newspapers and magazines, and have made their contribution to highlighting television news.

CHARACTERISTICS

A map must have the following characteristics if it is to be useful, regardless of the medium in which it appears:

- Accuracy
- Legibility
- Geographic perspective
- Visual impact
- Topographical detail when necessary

Accuracy

The design of a map often involves many time-consuming steps, such as locating an atlas or other similar source, digesting the message of the story the map accompanies, and deciding on color, size, and general design approach. However, these steps are generally wasted if the information on the map fails to pass that most important test of *accuracy.* Conscientious designers and artists do not underestimate even the smallest fraction of a distance to guarantee that a city or river appears where it's supposed to. Fortunately, maps used to accompany a news story are not expected to present the thousands of topographic details that one would find in an atlas. Even so, each map that is used in a news story must be prepared as accurately as possible. Because of deadline pressures, newspapers cannot handle the great number of details on a map that an atlas does, but an attempt is made to present readers with as accurate a map as possible. In television design, excessive detail would be lost on the screen, so accuracy is followed in terms of situation, surrounding areas, and the correct spelling of geographic designations. For example, when the graphic designer of a local television station misspelled Guantanamo, Nicaragua, and Afghanistan within *one* news segment, the credibility of the report suffered—and rightly so. What guarantee do the viewers have that a report is accurate if the maps accompanying the report contain misspelled words?

Professional cartographers at the National Geographic Cartographic Division often allow six months or more to prepare maps with guaranteed topographical accuracy. Graphic departments must have available such resources as Rand McNally, *National Geographic* magazine, Hammand maps, and U.S. Coast and Geodetic Survey maps as basic reference tools. In addition, the Columbia Gazetteer (a volume that is an encyclopedia of detailed information on cities, towns, and countries of the world) can be most helpful to verify location and spellings.

Legibility After ensuring the map's accuracy, making the material on a map *legible* is the graphic designer's second most important task. Television presents certain limitations in that a map is often used only in a small portion of the screen—usually about a quarter, which is a very reduced area to present such information. Another limiting factor is the short amount of *time* that the map will be on the screen. A final consideration is the size configuration of certain countries that may not adapt well to the 4 × 5 landscape ratio of the television screen.

Typographically, a good standard to follow is to use Helvetica Medium (upper case) for countries and states and lower case for cities. All water areas can then be presented in upper- and lower-case Helvetica Medium italic. Helvetica is a clean, easy-to-read typeface that is as clear on the screen as on a printed page. The same typeface should be used for all maps shown in a segment in order to maintain a certain pattern of continuity.

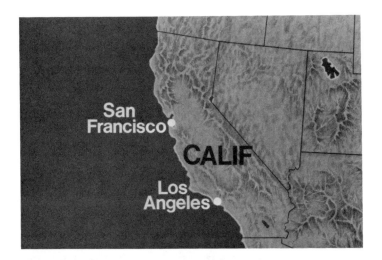

Frame 161. Typical map with topographical detail and major city locations.

Frame 162. For designers whose experience is primarily in the print media, a word of caution is in order: Typography on maps to be used "on screen," must be larger than that used on print maps.

Frame 163. An area highlighted by enlarging it, to present a reference point at a glance.

Because contrast is an important legibility factor, the most "relevant" type on a map should appear in white and the less important information presented in black. This, of course, is not a fast rule. Each designer will be flexible regarding type size and arrangement, depending on the information available and the message the map is supposed to convey.

In terms of placement, type can be applied to the map by using rub-off or transfer type, or it can be placed on an overlay (acetate) to avoid bruising the map. Type can also be superimposed from alphanumeric hardware.

For designers whose experience is primarily in the print media, a word of caution is in order: Typography on maps to be used on a screen must be larger than that used on a print map.

Geographic Perspective

The Vietnam War revolutionized the use of maps as informational graphics and, in the process, made viewers more aware of and dependent on topographical information. As the war became a daily part of the evening news presentation, so did maps of Vietnam. More recently, the same phenomenon has happened with maps of such troubled areas as the Middle East region, El Salvador, and the Falkland Islands.

What can the designer do to vary a design when an event lingers in the same geographic location and the "shape" of the area obviously remains visually the same? It takes creativity and inventiveness to bring about visual interest while using the same basic element, but it can be done—and it all begins with a sense of geographic perspective.

The following strategies deal with such perspective.

Frame 164. The focused area has the boldest color or shade, giving it a three-dimensional look.

A map used in an effective way will possess visual impact—it will command instant attention. Historically, maps have not been the most visually appealing of graphic tools. In television, maps have evolved from primarily flat, gray masses with borders defined in black and names of countries, cities, and so on in white and black to today's color maps of flat color

Visual Impact

Frame 165. *Above,* a map showing the relationship between the relevant area and its surrounding areas.

Frame 166. *Above,* the use of panning and zooming on a map. With countries such as Israel (in non-aspect ratio), this could effectively convey the idea of narrowness and that country's problems on various borders. Panning and zooming also enhance and dramatize the topographic information.

Frame 167. A limbo map isolates a country, city, or state from its physical environment; an edge is added around the borders to give it a new dimension. This technique is particularly useful in presenting areas that have been in the news for a long time and whose geographic locations are therefore familiar to the viewer.

Frame 168. The limbo map can be enhanced through the use of a symbol, flag, or other graphic device which creates a poster impact while conveying geographical information.

THE DESIGN AND USE OF MAPS AS INFORMATIONAL GRAPHICS   103

Frame 169. Showing routes on a map.

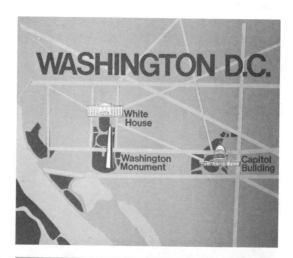

Frame 170. Highlighting the location of a disaster.

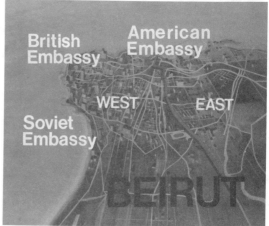

Frame 171. A close-up of Washington State highlighting by illustration the focus of the story on Mount St. Helens.

Frame 172. Indicating a trend.

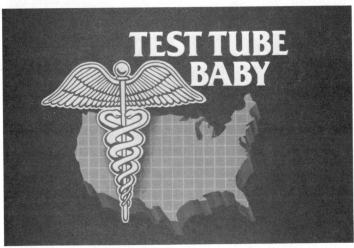

Frame 173. The caduceus is used frequently to give immediate identification to medical stories.

a. b.

Frame 174a–b. A map being prepared.

areas—usually yellow or different shades of tan against medium-blue water areas, with black or white typography.

Maps also gain visual impact by the addition of other elements such as indicators *showing routes* (such as a presidential motorcade, the evacuation of a city, or a military advance), *highlighting a catastrophe* (such as an explosion, a plane crash, or a train derailment), or *pointing to a trend* (such as the statistical spread of cancer in a section of the country, the homicide rate in a city or state, or unemployment figures). These indicators would probably be superimposed over the base artwork and not directly on the map.

Preparation Experienced designers know that the success of an accurate, visually appealing, legible, and geographically sound map depends on the amount of thought and preparation invested in its execution. Most experienced designers also agree that the key to preparation is the *anticipation* of needs by artists and those in charge of the day-to-day production of informational graphics.

Here are some tips on preparing maps and anticipating their use:

1. Start with base maps of major areas or continents.
2. Concentrate on a series of "limbo" maps of major trouble areas.
3. Prepare individual maps of major countries first as limbo maps, then with surrounding countries included.
4. Treat the United States as a complete map, then subdivide it according to sections; that is, east, south, southwest, west, northwest, and northeast. Prepare a map of each state separately, highlighting those important cities that are likely to become centers of news.
5. The best size for drawing maps ranges from 20 × 30″ to 30 × 40″. Slides and transparencies come in handy; they should be prepared in advance.

THE GRAPHICS OF MAJOR ANTICIPATED NEWS EVENTS

The challenge of "on deadline" creation of informational graphics for late-breaking news events allows the designer to apply creativity, ingenuity, and experience under the heaviest of pressure. Perhaps that is the reason why designers jump at the opportunity to become involved in the creation of graphics for major "anticipated" news events.

First, the nature of these events allows for planning, a luxury seldom found in the day-to-day production of late-breaking news. Along with planning comes ample time for designers to brainstorm, to experiment, and to eventually come up with graphics that will depart from the ordinary.

The following are the major anticipated news events with which we will deal in this chapter:

- Election campaigns/election coverage
- Space-related coverage
- Olympic games
- Other events

To the visual graphic designer, these are long-term projects, many of which are planned in advance as much as one year, allowing designers and those involved in the journalistic aspects of the coverage to meet, exchange ideas, and plan the best strategies for making the content easier to understand.

ELECTIONS

Coverage of an election, especially one of national prominence, begins one year prior to its scheduled date. The first step is to create and implement a logo design that can then be used as a symbolic identifier to link the viewer and the station. Logos emphasize a color scheme that, when accompanied by a representative, legible typeface and illustration, creates a "memorable" symbol.

But visual coverage of an election is more complex than simply featuring a symbolic logo. The essence of any election involves *the people, the issues,* and, of course, *the results.* Photographs of all candidates are secured and prepared, along with the candidates' background data and their stand on the issues.

An important national election also involves geographical breakdowns, including such sociological variables as vote by different ethnic groups, urban versus rural, and young versus old. Designers, working closely with producers and researchers, can prepare charts accordingly.

Continuity and consistency in the use of color, type, and photographs become key factors in organizing and maintaining effective graphic coverage of a major event. As soon as the designer has made choices to standardize those areas, the rest of the task is simplified, allowing more time to verify the accuracy of the material.

On any of these longer-term projects, manual graphics are kept to a minimum or are eliminated entirely. These graphics are produced on graphic display devices such as the Dubner Color Background Generator (CBG), Quantel Paint Box, and Chyron, and are stored for retrieval on discs and videotape. (See Chapter 9.) Designers work very closely with the producers and directors on these projects. As soon as the original design ideas are approved by everyone, then the work can be put on computer graphics machines.

The CBG operator can work from black-and-white drawings of maps or other designs in order to digitize and color. The machine allows for modifications to be made until an effective design is achieved. During the coverage of an election, for example, state, regional, and national maps are prepared and color-coded to attach meaning to the coverage. Special effects—such as having a state map zoom out of a national map—can be achieved. Or a photo of a winning candidate surrounded by data and a bar chart showing figures connected with his victory can be prepared as well.

Electronic devices have had a great impact on the preparation of informational graphics. For elections, the opening animations and bumpers (a bumper is a visual that separates the show from the commercial, and can be a show title, a still frame from the show, or some art) are usually a combination of

Frame 175. The logo for the 1980 election campaign coverage for ABC News.

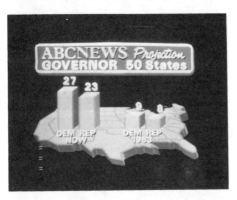

a.

b.

c.

Frame 176a–c. Nationwide election results for governors and Senate and House races done as bar charts.

Frames 177–178. Election coverage facilities in the 1960s.

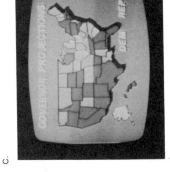

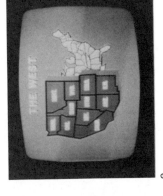

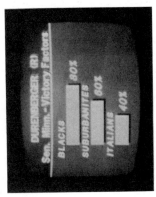

Frame 179a–g. Examples of ABC News election coverage graphics in the 1970s.

electronic graphic devices and/or film and tape animations, using components of the overall logo designs. These animations are prepared for a full screen and are superimposed over the electron set.

Before the advancement of technology, it was not always so easy to prepare the informational graphics that accompany the coverage of an election. The progression in graphic techniques has been an interesting and remarkable one, from physical studio boards with handwritten numbers in the 1950s, to slotted boards for insertion of numbered cards in the 1960s, leading to the mechanical digital display "solari" numbers of the 1970s—similar to those in digital clocks. During these transitional periods from actual studio boards to electronic boards, all networks backed up the electronic boards with the manual, not wanting to risk the perils of a breakdown—especially for an event where accuracy in numbers could make the difference between a winning and a losing candidate.

Accuracy and clarity of type are the two most important factors in preparing boards. Designers and their assistants spend considerable time verifying information, knowing the impact that such data carries. Presenting a seemingly boring mass of numbers, names, and percentages does not have to be a painful experience for either the designer or the viewer. Electronic devices are so sophisticated today that a candidate's picture can be shown along with some animation. In addition, shaded, air brush-looking backgrounds, metallic effects, and various textures and dimensions can also be achieved.

Sets, too, become part of the overall graphic package of a major anticipated news event. The illustration below shows how the 1980 ABC set included an enlarged version of a logo that was used as a visual representation of that event.

SPACE SHOTS

Few designers can argue the fact that coverage of the space program presents the most graphic possibilities of any other event of our time. From the inception of the space program in the 1950s, television has capitalized on it as the "graphic" event that it is. Here is an event in which the terms that refer to the subject are graphic descriptions: launching, splash down, space walk. The subject is fascinating and affords designers the freedom to experiment. In the 1950s, there was no facility to have vehicles in space and shoot additional photos after a spacecraft was launched. The responsibility of making inter-

a.

b.

c.

d.

Frame 180a–f. Photos of ABC election coverage set model.

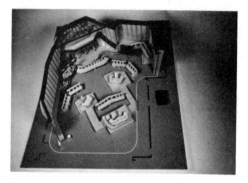

e.

f.

esting, informative shots fell in large part on graphics and special effects.

Creativity and perseverance came to the rescue, then, often creating visual clarification of events in space that are not clearly covered by a camera.

Today creativity and perseverance continue, but are greatly assisted by computerized graphics—and the coverage made possible through photographs taken in space by the astronauts. Obviously, the material continues to be fascinating, with the element of the "unknown" the main ingredient that keeps graphic designers busy—and millions of viewers interested.

Planning for a shuttle launch begins almost immediately after the previous mission. Each mission program is given special treatment and planning for graphic illustration. Layouts are designed to include live picture, elapsed-time digital clock, and a location map indicating at all times where the shuttle is in relation to the earth.

Landing-site maps and maneuvers are graphically illustrated to keep viewers informed about other phases of the mission. Each space mission is given its own visual identity through a logo that has typographic continuity for coverage of the entire event. Most of the graphics are put on the Dubner Color Background Generator. In the case of orbital maps, an art card map is used in conjunction with electronic orbit route overlays, including the shuttle symbol. The maps are then videotaped and stored for use in shuttle on-air programs.

OLYMPIC GAMES Preparing graphic coverage of the Olympic games requires a great deal of organization along with creative ways of presenting the same material in different ways. Usually more than a year is dedicated to working on the specifics of graphic symbols, logos, and other materials that will be used in conjunction with coverage of the games.

Fonts and logos are stored in the Dubner and Chyron machines; other materials are placed in art card form.

Such symbols as flags, names of sports, and names of participants are prepared in advance, allowing for design and style to blend into a consistent pattern of visual organization. This process of organization leads to easier production during coverage of the event.

PRODUCTION TIPS FOR THE VIDEO GRAPHIC DESIGNER

So far we have treated the most aesthetic aspects of design, concentrating on results for an appealing graphic presentation of information. In this chapter we turn our attention to the most practical aspects of producing an actual frame, emphasizing the most direct and economical approach in each case.

The success of any of the following techniques relies heavily on the graphic director's ability to use personnel to the best advantage. What is a tedious task to one artist may be an exciting challenge to another. An interlocking relationship can be effected for consistency of style, allowing artists the flexibility to concentrate on those techniques that they perform best. Efficiency of personnel is one of the key components of good graphics, an area that involves management strategy more than the artistic aspects of production.

The following frames demonstrate the simple steps required to achieve the most direct and economical approach to design problems.

Frame 181. A basic map can be prepared as a card with a silkscreen paper prepared to specifications. It is similar to Color Aid or Color View paper, and can be utilized for both land (tan) and sea (blue) values.

Frame 182. The first step in the preparation of a map is to do a pencil tracing from a source map.

Frames 183–184. The tracing is transferred (usually enlarging it by statting it from source material) to Color Aid paper. The land area cutout is then sprayed with rubber cement and is pasted on to the blue sea area paper, which has previously been applied to an illustration board.

Frames 185–186. With a color pencil, the essential elements (mountains, terrain, etc.) are put on the map and the necessary borders defined with a Radiograph pen or brush or Pentel, maintaining a consistency of line.

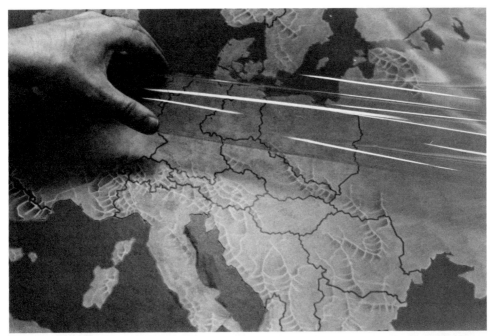

Frame 187. The map is covered with acetate, which is taped to the back of the board. All additional information is put on the acetate cell in transfer type of various sizes and in black and white, depending on the value of the background. This way type can be quickly removed or added without damaging the basic map.

Frame 188. If a country needs to be highlighted, a lighter piece of colored paper can be placed over the acetate with a drawn dimension on the map to give the optical effect of the country being raised from the surface.

Frames 189–190. A limbo map can be prepared from Color Aid paper in the same way as a regular map. The map can be identified with rub-off type (sometimes applied directly to the map) and is usually placed on a black background. This is always done with dimensional rendering.

Frames 191–192. Not all maps are run as individual pieces of art; often, they accompany head shots, flags, a globe, corporate logos, an oil barrel, or just typography or a live picture. If any of these additional elements are used, they are placed on separate cards with chromakey background, usually green (but of any color, including black, as long as the key color is not in the art itself). Then the artist combines the map with other elements singly or in any fashion desired.

193. Graphic statement for unemployment.

COMPUTER-GENERATED GRAPHICS AND THE MACHINES THAT PRODUCE THEM

COMPUTER-GENERATED GRAPHICS

The most important impact on graphics in recent years—and one that is revolutionizing the graphics product on television—is the appearance of computer graphics machinery that has the capability to produce the most sophisticated graphics with animation. Ironically, computer-generated graphics originated within the television medium itself. The presence of computer-generated graphics goes beyond regular television news and other programming to expand into the new but vibrant field of cable-delivered informational services.

This chapter outlines what is available in the technology of computer-generated graphics, the new possibilities allowed to artists and designers, and the impact their choices will have on production. But there is a very frustrating element to cope with when dealing with such subjects, and that is the fact that the state of the art in computer-generated graphics changes very rapidly. To that effect, our attempt is to describe what is available in the mid-1980s, reassured that the basic principles of computer-assisted graphics are common to all systems, perhaps present and future.

Nowhere is computer-generated graphics more popular and effective than in the production of news graphics. Demanding deadlines and the ever-changing nature of news events create the ideal framework for digital systems. News artists can save tremendous time in the process of conceptualizing and executing ideas. Robert Walkens, a producer and computer artist, described in an article in *Computer Graphics World* (June 1983) how a news artist was asked to create a graphic for Chrysler Corporation using the usual generic car or fact graphic. "The artist sweated through a half hour of searching for ideas before realizing that his car key had the Chrysler logo stamped on it. Placing the key under a video camera with a line in to the paint system, he digitized and enhanced the logo, quickly created a high-tech background, and added a title with the font mode."

The price of a computer that generates graphics ranges from a few thousand dollars to more than a million dollars, depending on the degree of resolution and capability it offers. Electronic graphic displays have been on the market for twenty years. Early forms were used for designing cars, aircraft, and in-flight training; in the past five years systems have become versatile and easier to use. This has had a big impact on the television industry, specifically in the categories of information and design. Electronic graphic displays are used for in-house promotion animation and for sports and news programming.

What are the advantages of computer-generated graphics over manual graphics? Why is an investment in electronic graphics equipment sound?

1. The primary advantage is speed. Whatever the storage device is, the almost instant retrievability is a major plus.
2. The direct electronics image of the graphics display is a cleaner, sharper, brighter picture than that fed from a card to a camera to the tube.
3. The wear and tear on manual graphics is eliminated.
4. Integration with other electronic elements is easier and faster.
5. The storage requirements are smaller than for manual graphics.
6. The availability of equipment at all times (if the machinery is properly organized) eliminates the need for calling in additional staff or creating a graphics department.
7. The new graphic devices afford a variety of graphic styles at the fingertips of a good operator. Producing a similar design manually would require the talents of the most superior graphic artist.
8. With sufficient storage capability and anticipation, many graphics can be programmed and stored in advance.
9. Another advantage is in clearing the studio floor of graphics cameras, cameramen, easels, and storage room—in many cases even eliminating studios.
10. Using electronic graphics equipment makes many laborious graphics automatic, such as putting maps or lettering in perspective; masking and silhouetting; changing colors in a

given area; and changing any element in a design or repeating a pattern or design. In fact, an artist can generate four times as much artwork with a computer graphics system, producing images far superior in every way.

The following is a list of the best-known computer-generated systems.

1. *AURORA/100.* Described as one of the fastest graphics idea machines in terms of daily broadcast use, it has normal tablet and digital input devices with real-time animation effects. It has cut-and-paste, snip, tear, throwaway, paint, wash, rule, size, and multiply commands, and shadow and typography modes. It is a very fast system. For recording on video or film, a special menu is incorporated to save production time by keeping a record of each frame as it is recorded.
2. *BOSCH FGC 400.* Produces an impressive picture with continuous shadow by automatic light source for a real three-dimensional look. It is expensive, needs long lead time, and is not easy to operate.
3. *CHYRON.* The workhorse of the industry is Chyron IV. It is probably the best and most reliable of the typography systems. It includes a graphics component package with a multimode graphics module and digiflex special effects, offering an outstanding variety of quality video typefaces and 512 colors, 64 of which can be on the screen at one time. The Chyron system offers a very good and reliable graphics display device, with "tricks" not found in any other system in its moderate price range.
4. *COMPUTER GRAPHICS LAB.* Conventional print system effects and design with real-time animation. It has a good look of depth and gradation of color, and a variety of airbrush, repeat, and cut-and-paste modes make this a good resolution, user-friendly, moderately priced system. One mode allows the user to paint through a precreated twelve-field graphic that is viewed on a six-field frame, with an overlay animation running independently.
5. *DUBNER COMPUTER SYSTEMS (CBG 2).* This system has paint mode and font generation, and real-time animation. Picture quality is excellent but does not have the gradation of tone of some other devices. An easy add-on approach makes this system an easily upgradable, flexible system that is always improving. It has good resolution. The CBG has 4096 colors, 128 of which can be on the screen at one time. It is the only vector graphics three-dimensional animation device. Its digitizing quality is not first-rate, but improvements are on-line.
6. *MCI QUANTEL.* Referred to as the Paint Box system, it is fast becoming the most popular of all systems—and for good reason. It is also the most user-friendly system, with

Frame 194. Election graphics produced entirely from the Paint Box.

Frame 195. Wall Street generic graph from the Paint Box.

Frame 196. The American flag background design produced by the Paint Box.

Frame 197. Sports illustration using the Paint Box.

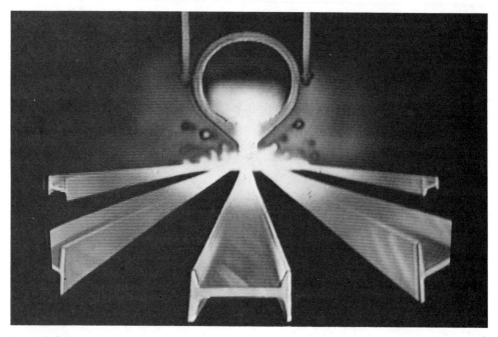

Frame 198. Generic frame for the steel industry.

Frame 199. Basic map of Texas and accompanying illustration, totally executed on the Paint Box.

Frame 200. Special graphic prepared for a report on cable television.

Frame 201. Introductory graphic for a baseball score designed and executed on the Paint Box.

the closest duplication of the artist's tools. Airbrush and frisket are excellent, along with cut-and-paste, and repeat modes and other characteristics described below.

Digitizing is superb, with 100 percent fidelity, and users describe the retouching ability as brilliant. The following characteristics are what make the MCI Quantel a prominent machine in today's industry.

a) Special animations in real-time are possible, if used with a frame storage system. However, it is not a complete real-time animation.
b) Ability to take in live video feed—allowing changes to be made with the feed, then be given back out.
c) A complete graphics studio for the development of single-frame video.
d) Stunning perspective designs can be achieved with art and type when used in conjunction with the Ampex ADO.
e) It uses all the traditional skills employed by an artist, replacing only traditional materials.
f) Helps graphic designers produce 5 to 10 percent more work at higher standards in far less time than traditional tools.
g) Higher resolution, providing more vivid color than that on art cards.
h) Variations of colors to more than six million, with the use of airbrush, chalks, points, and different-size brushes.
i) Type fonts that are sharp and clearer than what most other machines can produce (limitation of only one line at a time). In addition, 150 typefaces are available through the system.

7. *QUANTEL "MIRAGE."* This is a highly sophisticated video manipulation and compression system, that can best be described as a video origami (paper-folding) machine. It takes two-dimensional art or a picture, and creates three-dimensional video. For example, a flat Mercator map can be turned into an opaque, translucent, or transparent three-dimensional globe.

8. *ADDA.* This is an add-on graphics system used at Cable News Network. Reliable, user-friendly equipment, it has the ability to show you results and allow you to add on as you proceed. The add-on can be up to twenty generations without appreciable loss of quality. Good switcher and stiff store with compliment of airbrush and cut-and-paste, softedge, wipe, and other modes.

9. *VIDIFONT.* A user-friendly system similar in many ways to the Aurora system. Its drawback is a cumbersome digitizing approach that cannot take a full screen image at one time but, instead, must be executed in quadrants of the screen. It offers good typography and includes 512 colors.

10. *G. E. GENEGRAPHICS.* This is one of the pioneer systems that is equipped with a vast file of graphics symbols and drawings. A more complete software package is presently being developed.

This has been only a brief review of the more popular systems used in television production, with the main thrust in news graphics production. There are, in addition, many special effects devices available at the control-room level that can aid in visual presentation, including the following:

    a) The single-channel or multiple-channel Ampex ADO, which can rotate, tumble, and put in perspective or angle full video or typography on any portion of the screen with high resolution.

    b) The Quantel plus autoflex mode with multiple channels, which can distort, rotate, angle, and perspective any video input with five-channel capability.

    c) The Quantel 3100, which can achieve much of the same effects on three separate effects decks.

One should explore all systems before deciding to purchase one. It is probable that the growth of the field will level off and the strongest system or systems will emerge to dominate the market. As electronic graphics development has provided television directors and producers with the additional tools to make their product more interesting, it has also given television stations the ability to do more exciting and visually effective promos and program titles. Most importantly, the technology has relieved the graphics department of much of the drudgery of creating graphics.

One word of caution is in order: Artists should beware not to sacrifice originality, simplicity, and directness for efficiency. Each computer has certain characteristics—color palette, airbrush modes, and so forth—which give it an identity, and there is a tendency for graphics to start looking too uniform if they are not laid out before execution. There is the inherent danger of overusing the machine's features (tumble, zooms, perspectives, and so on), which could create a sameness that one was presumably looking to eliminate by purchasing the new equipment.

In creative hands, however, the equipment can perform with magnificent results in record production time. Special vigilance must be maintained for the other side of that coin—elaborate failures.

Computer-generated graphics have also become part of an "electronic publishing" revolution that is dramatically changing the way in which traditional newspaper content becomes accessible to the public. Publishing via the home television set—commonly referred to as *Videotext*—is growing rapidly through the proliferation of cable television systems.

The videotext industry has three technological variations, which are all dependent on some form of graphic device to attract and hold the attention of the viewer:

1. Rolling page
2. Teletext
3. Videotext

At the Sentinel Communications Company in Orlando, Florida, the rolling-page concept developed as a direct outgrowth during the planning stages of producing a teletext service. The result was the creation of *News Sketch,* a graphically oriented newspaper on the air which is described as a headline service, although it also offers advertisement and bulletin board-type announcements. News Sketch, a twenty-four-hour, seven-day-a-week electronic information service, began on November 10, 1982, on Orange-Seminole Cablevision in Orlando. It has been fully operational since that time.

This service, which was originally based on the Canadian Telidon standard and later converted to the North American Presentation Level Protocol Standard created by AT&T, makes use of alphageometric graphics. These graphics are created by using a mixture of geometric shapes, including the circle, rectangle, arc, and polygon. The most commonly used graphic in complex artwork is the irregular polygon.

This system, built on a screen matrix of 200 lines horizontally by 256 vertically, offers a much finer resolution than is available on other systems. Many systems, such as the British Prestel and Ceefax, use alphamosaic graphics in which rectangular blocks of color are combined into graphic elements, very much like mosaic tile.

Creation of the graphics is accomplished using four Norpak IPS-2 terminals. These are manufactured in Canada, and are specifically designed for producing graphics. Each artist is equipped with a magnetic tablet and stylus, allowing the artist to draw in much the same way as he or she would with a pencil. Additionally, the cursor on the Norpak screen can be manipulated from the keyboard for doing very fine work one pixel at a time.

Output of the graphics and text to the cable channel is digital in nature; information about the picture is sent out one bit at a time. A Norpak Mark 4 decoder at the cablehead end decodes the digital transmission and provides video for the cable channel.

The time involved in transmitting and building pictures on

the screen adds another dimension to the work of artists. Artists must be concerned not only with the appearance of the finished art but also with the order in which elements of that art are allowed to appear on the screen and the time it takes to build the entire picture.

This time factor has some definite effects on the design of pages. For example, pages with both art and text must build quickly enough so that the reader will have time to read and understand the message. If the graphic content takes too long, the page may be replaced before the message can be read.

Colors are another key factor in design. Although in theory 32,000 colors are available to the artist, only 16 may be used in any one picture. This is a limitation of the decoder, which can remember only 16 colors. Experimentation has also shown that some colors simply are not decoded well, and must be used with care. The biggest offenders are purple and red, which tend to smear on the screen when used in large areas.

With the increasing demands for new effects for television, the development of electronic graphics can be a force for clarifying, enlightening, informing, entertaining, and stimulating viewers. Computer graphics have already had a big impact in print and typographic design, and will be an increasingly important element in creating better television. In addition, the further development of computer graphics will inevitably lead to some convergence of science and art.

The frames on the following pages have been videogenerated through the use of one or more of the systems described in this chapter.

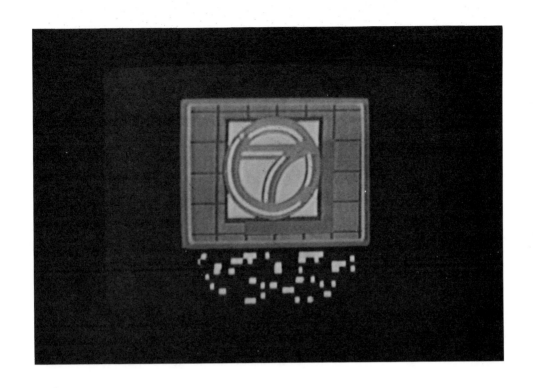

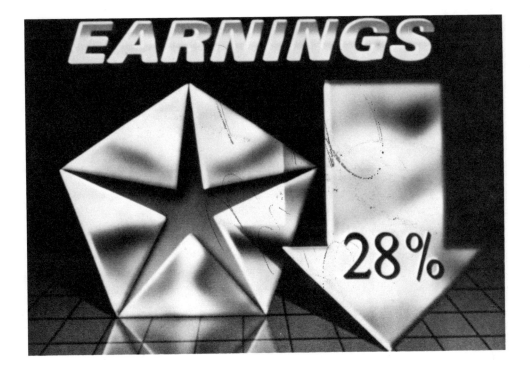

The frames on the following pages were prepared by the staff of the Sentinel Communications System to show what their videotext system is capable of doing.

Olympics

Former Pres Jimmy Carter said it ``would be a serious mistake'' to block the Soviet Union from the '84 Games.

Carter orchestrated the US boycott of the '80 Moscow Summer Olympics after the USSR invasion of Afghanistan.

Hockey

Wednesday NHL games

Hartford at Buffalo

New Jersey at NY Rangers

St Louis at Chicago

Boating forecast

Atlantic Coast:
Winds: S 10-15
Seas: 2'-4'

Gulf Coast:
Winds: SW 10-15
Seas: 2'-4'

World/National

   Lech Walesa said Wednesday he will ask a relative to pick up his '83 Nobel Peace Prize and will give the money to Poland's Roman Catholic Church.

   Walesa, founder of the now-outlawed Solidarity labor union, won the prize today.

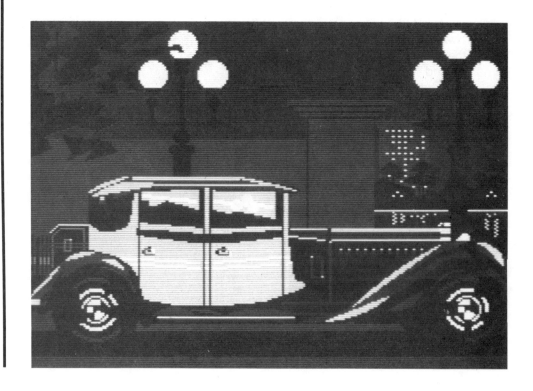

ABOUT THE AUTHORS

BEN BLANK was born in San Francisco in 1921 and graduated from Cooper Union. He joined CBS News in 1951 and became Graphics Director in 1952, a position he held until 1962. From then until 1985 he served as Graphics Director at ABC News. He is now the Managing Director of the new ABC Broadcast Graphics Department, which combines the former departments of News Sports and Entertainment Graphics.

He developed the concept of generic graphics for continuing news stories in the late 1960s. He is generally credited with much of the development of graphics in the television industry and is acknowledged to be the "Father of Television Graphics."

MARIO GARCIA was born in Cuba in 1947 and came to this country at the age of fourteen. He received a Bachelor's degree from the University of South Florida, with Master's and Doctoral degrees from the University of Miami. He is an Associate Director at the Poynter Institute for Media Studies in St. Petersburg, Florida, where he directs the Graphics/Design Center. In addition, he serves as professor of Mass Communications at the University of South Florida and as visiting professor of Graphic Arts at Syracuse University's Newhouse School of Public Communications.

Dr. Garcia spends part of his time serving as a consultant to newspapers across the United States, Canada, South America, and Europe. He is the author of *Contemporary Newspaper Design: A Structural Approach* (Prentice-Hall, 1981). In 1983 he was installed in the Scholastic Journalism Hall of Fame at the University of Oklahoma's H. H. Herbert School of Journalism and Mass Communications, and in 1984 he received a Fulbright Fellowship to lecturer on newspaper design at the Center for Professional Training of Journalists in Quito, Ecuador.

WITHDRAWN